The Philosophy of Veritism

The Philosophy of Veritism
Copyright © 2025
Rob Taylor Books Publishing
Rob Taylor Author

Cover Art by robtaylorbooks.com
Cover Copyright ©Rob Taylor Books Publishing
Rob Taylor Author

Rob Taylor Books Publishing and Rob Taylor Author supports the right to free expression and the value of copyright. The purpose of copyright is to encourage writers and artists to produce creative works that enrich our culture.

Unauthorized distribution of this book is a violation of the author's intellectual property. If you would like permission to use material from this book (other than for review purposes), please contact Rob Taylor Books Publishing robtaylorbooks.com
Thank you for respecting the author's rights.

Ebook: ISBN: 978-1-0698264-7-3
Paperback ISBN: 978-1-0698264-1-1
Hardcover ISBN: 978-1-0698264-4-2

Author's Note: On the Word Veritism

Words matter. They carry history, implication, and potential. When choosing a name for this philosophy, I wanted something that reflected its essence: truth grounded in reality, tested by evidence, guided by reason, and lived daily.

The word Veritism comes from the Latin veritas — truth. Historically, when the term has appeared, it has been associated with abstract discussions of truth, or the tension between truth and understanding. But it has never been firmly tied to reality itself.

That is where this work departs from the past.

In this book, Veritism is not simply "the study of truth." It is the philosophy of alignment with reality. Truth is not an abstraction here — it is correspondence to what is. Evidence is not optional — it is the measure of truth. Reason is not secondary — it is the tool of comprehension. And life itself is not detached from these — it is to be lived in harmony with reality, not illusion.

This deliberate reframing gives Veritism its unique character. Where religion demands faith without evidence, Veritism demands evidence before belief. Where atheism rejects gods but leaves a vacuum, Veritism affirms reality as the foundation for a positive philosophy of life. Where relativism dissolves truth into opinion, Veritism anchors it in fact.

By naming this philosophy Veritism, I am not inventing reality — reality needs no invention. I am naming what has always been present but too often ignored: the path of truth through reality.

I offer this note so that future readers understand: Veritism is not borrowed whole, but redefined deliberately. It is not the repetition of an old concept, but the creation of a clear, practical system that can guide us through illusion into truth.

The Philosophy of Veritism

Veritism — from veritas (truth): the way of truth

A Manifesto of Reality, Facts, and Evidence

Contents

Introduction

Preface

- Why this philosophy is necessary.
- The inadequacy of religion, atheism, and relativism.
- A call for a system grounded in base reality.

Part I: The Foundations of Reality

Chapter 1 — Reality Exists

- The first recognition.
- The cost of denying reality.
- The first integrity.
- The first principle of Veritism.

Chapter 2 — The Nature of Facts

- What is a fact.
- Fact vs opinion.
- Why facts matter.
- The second principle of Veritism.

Chapter 3 — Truth as Correspondence

- What is truth?
- Belief vs truth.
- The ethical dimension of truth.
- The third principle of Veritism.

Part II: Knowledge, Reason and Evidence

Chapter 4 — Evidence: The Measure of Truth

- What is evidence?
- The hierarchy of evidence.
- The abuse of evidence.
- The fourth principle of Veritism.

Chapter 5 — Reason: The Tool Of Comprehension

- What is reason?
- Reason vs emotion.
- The discipline of rational thinking.
- The fifth principle of Veritism.

Chapter 6 — Knowledge: The Integration of Truth

- What is knowledge?
- The difference between knowledge and opinion.
- Knowledge as power and responsibility.
- The sixth principle of Veritism.

Part III: Wisdom, Ethics and Freedom

Chapter 7 — Wisdom: The Application of Knowledge

- What is wisdom?
- The practical nature of wisdom.
- Collective wisdom.
- The seventh principle of Veritism.

Chapter 8 — Ethics: Living in Alignment with Reality

- What is ethics?
- The false roots of ethics.
- Social ethics.
- The eighth principle of Veritism.

Chapter 9 — Freedom: The Power to Live by Truth

- The essence of freedom.
- Freedom and truth.
- The fragility of freedom.
- The fragility of freedom.
- The ninth principle of Veritism.

Part IV: Purpose Living in the Real World

Chapter 10 — Purpose: The Meaning of Life in a Real World

- What is purpose?
- Purpose and reality.
- Purpose and mortality.
- The tenth principle of Veritism.

Chapter 11 — Community: Shared Reality and Collective Flourishing

- Shared reality as the basis of community.
- Community and ethics.
- The fragility of community.
- The eleventh principle of Veritism.

Chapter 12 — Fulfillment: Living the Whole of Truth

- The path to fulfillment.
- Fulfillment and freedom.
- The joy of fulfillment.
- The twelfth principle of Veritism.

Part V: Legacy in Transcendence

Chapter 13 — Legacy: Truth Across Generations

- The nature of legacy.
- Legacy through ethics.
- Humanity's legacy.
- The thirteenth principle of Veritism.

Chapter 14 — Transcendence: Truth Beyond the Self

- Reality as the ultimate context.
- Transcendence and legacy.
- Mortality and transcendence.
- The fourteenth principle of Veritism.

Chapter 15 — Practice: Living Veritism Daily

- The discipline of awareness.
- Reason as daily compass.
- Facing illusions and lies.
- The fifteenth principle of Veritism.

Conclusion — The Call of Veritism

- The simplicity of reality.
- The integration of life.
- The future of truth.

Closing Words

Acknowledgments

Preface

Every age faces its illusions. Some are religious, promising eternal life without evidence. Others are ideological, offering perfect societies on the basis of slogans. Still others are personal, the quiet lies we tell ourselves to avoid discomfort. Yet all illusions share one thing in common: they separate us from reality.

For centuries, religion has bound humanity to unverifiable claims, demanding faith without evidence. In response, atheism has grown, rejecting gods but too often leaving a void — a framework defined more by denial than by affirmation. Relativism, in turn, has whispered that truth is subjective, that each person may shape reality as they please. But if truth dissolves into opinion, then nothing can be distinguished from illusion at all.

Each of these paths is inadequate. Religion offers dogma in place of fact. Atheism stops at negation without building. Relativism erodes the very ground of truth. What we need is not faith, not rejection, not dissolution — but alignment. Alignment with the one thing that stands independent of belief or denial: reality itself.

This book was born out of a simple conviction: that life must be lived in alignment with what is real. If reality is the foundation, then truth is its recognition, evidence is its measure, and reason is its tool. Without these, we are blind. With them, we can see clearly, act wisely, and live fully.

I have called this system Veritism, from the Latin veritas — truth. It is not a religion, for it demands no faith. It is not an ideology, for it prescribes no utopia. It is not atheism, for it is not defined by rejection, but by affirmation: the affirmation that reality exists and is knowable. Veritism is a philosophy of alignment — alignment of thought, action, and life with the undeniable fabric of reality.

The aim of this book is twofold:

1. To provide a clear framework for understanding truth in all its dimensions.

2. To offer practical guidance for living daily in harmony with reality.

The chapters are structured as steps in this journey: from the foundation of reality, through the nature of facts and truth, into evidence, reason, knowledge, and wisdom, and then into ethics, freedom, purpose, community, fulfillment, legacy, transcendence, and practice. Together, these fifteen principles form a compass — not a map, for no map can capture the vastness of reality, but a guide that points always toward what is true.

This work does not ask you to believe. It asks you to see. It asks you to test, to measure, to reason, to live.

If you find clarity in these pages, it will not be because of persuasion, but because reality itself is persuasive. If you find strength here, it will not be given by me, but by truth itself.

Veritism is not mine to own; it belongs to reality. It is simply a name for what has always been there: the path of truth.

This book is my invitation to you — to walk that path, to reject illusion, and to live fully in alignment with what is.

Part I: The Foundations of Reality

Chapter 1 — Reality Exists

Reality is the one truth that requires no defense, and yet it is the one truth most often denied. Every superstition, every ideology, every unfounded belief stands as testimony to humanity's uneasy relationship with what is. We build temples to gods, but no temple to gravity. We tell stories about divine creation, but no story about the fact that two plus two equals four.

And yet, beneath all illusion, reality remains.

1.1 The First Recognition

Every philosophy, every worldview, begins with a starting point. For some, it is faith in a god. For others, it is belief in the authority of human reason. For Veritism, the starting point is simpler and more unshakable: Reality exists.

This recognition is not belief. It requires no leap of faith, no acceptance of dogma. It is the recognition that the universe does not depend on us to be real. Whether we acknowledge it or not, the sun will rise tomorrow. Whether we accept it or not, we will one day die.

The first act of integrity is to face this without denial.

1.2 The Independence of Reality

Reality does not bend to human will. A culture may believe the earth is flat, but ships will still vanish over the horizon. A man may insist he can walk through fire unburned, but the flames will not negotiate.

This independence is what makes reality trustworthy. It does not conform to opinion; it stands apart. That which is real remains real regardless of what we think about it.

If reality were created by thought, truth would shift endlessly, knowledge would dissolve into chaos, and no progress would ever be possible. The fact that we can build bridges, cure diseases, and launch satellites proves that reality is stable — not a reflection of belief but a foundation beneath it.

1.3 Illusion and Existence

Human history is a long struggle between illusion and existence. We are creatures capable of imagination, and imagination often overreaches into claims of truth.

- Ancient people imagined that thunder was the anger of gods. Reality revealed it to be atmospheric discharge.

- Medieval healers imagined that disease was punishment for sin. Reality revealed it to be caused by microbes.

- Some today imagine that climate change is a hoax; others imagine it as an apocalypse. Reality, however, is neither hoax nor hysteria. The Earth's climate has always shifted — through ice ages, warming periods, and natural cycles written into the planet's history. What is real is change itself: coastlines rise and fall, weather patterns shift, ecosystems adapt or collapse. What is not real are the illusions that governments or ideologies spin — promising salvation or doom to serve their own agendas.

Illusion comforts, but it does not cure. Existence heals, builds, and reveals.

To live by illusion is to build on sand. To live by existence is to build on stone.

1.4 The Cost of Denying Reality

The denial of reality always carries a cost. It may come slowly, but it comes.

When leaders deny truth, nations collapse.

When doctors deny facts, patients die.

When individuals deny reality, they suffer, even as they cling to their illusions.

Consider history:

- In the Middle Ages, plagues swept through populations that rejected evidence in favor of superstition.

- In the 20th century, millions died under regimes that placed ideology above reality.

- In the 21st century, people die when facts about health, climate, or technology are ignored.

- Reality is not cruel. It does not punish us for ignorance. It simply continues — and we suffer the consequences of denial.

1.5 The Dreamer and the Sleeper

There are two ways to live: asleep in illusion or awake in reality.

The dreamer invents a world of his own making. He believes the universe is designed for him, that his wishes shape existence, that what feels good must be true. But when he acts, reality unmasks him. His prayers do not stop bullets, his beliefs do not cure disease, his illusions do not feed him.

The one who wakes sees reality as it is. He does not demand it be otherwise. He builds his life on facts, not wishes. He does not control the universe, but he controls his response to it. He cannot command reality, but he can align himself with it — and in doing so, he thrives.

1.6 The First Integrity

To acknowledge reality is the first integrity. Every lie begins with a refusal to see what is. Every delusion begins with the decision to substitute belief for fact. Every tyranny begins with the denial of truth.

The first act of honesty is not toward other people but toward existence itself. It is the recognition that reality does not require permission to be real.

Integrity means this: I will not deny what is.

1.7 Counterarguments and Their Collapse

Some claim that reality is an illusion. Philosophers have long debated whether the world is real or merely a projection of the mind. But notice this: even the argument that "reality is illusion" requires a real mind in a real place having the argument. The denial of reality collapses into contradiction.

Others claim that reality is created by belief. But if this were true, then two people with opposing beliefs would create two opposing realities. Yet when they both step into fire, both are burned. Reality answers not to belief but to evidence.

Still others argue that reality is unknowable. While it is true that no human has absolute knowledge of everything, it does not follow that we have no knowledge at all. We may not know the full nature of the universe, but we know enough to build houses that do not collapse, planes that fly, and medicines that heal. Partial knowledge does not negate reality; it affirms it.

1.8 The Unwavering Ground

Reality is not our enemy. It is the only ground we have. Every human achievement — from the wheel to the internet, from the alphabet to space travel

— has been possible only because reality is consistent and discoverable.

The universe is not cruel, nor kind, nor personal. It simply is. But within it lies the opportunity for truth, meaning, and progress. The man who aligns with reality is strong. The man who denies it is already broken.

1.9 The First Principle of Veritism

Thus we arrive at the first principle of this philosophy: Reality exists.

This is the cornerstone of Veritism. Upon it rests every other truth: facts as the measure of truth, evidence as the test of knowledge, and integrity as the proper response to existence.

To begin anywhere else is to begin in error. To deny it is to dissolve into illusion. To embrace it is to stand on the only ground that cannot be shaken.

Chapter 2 — The Nature of Facts

Facts are the language of reality. They are the bridge between existence and understanding, between the world as it is and the mind that seeks to grasp it. Without facts, thought becomes speculation, belief becomes illusion, and knowledge dissolves into opinion.

To live in truth, we must first know what truth is made of.

2.1 What Is a Fact?

A fact is a statement about reality that can be demonstrated to correspond with what exists.

- "The Earth orbits the Sun" is a fact, demonstrated through observation and measurement.

- "Water boils at 100°C under standard atmospheric pressure" is a fact, demonstrable through repetition.

- "I was born" is a fact, verifiable by record, memory, and physical existence.

A fact is not created by belief. It is not altered by denial. It is not strengthened by agreement. A fact simply is — its authority derives from reality itself.

2.2 Fact vs. Belief

A belief is an internal conviction. A fact is an external demonstration.

- A person may believe that praying for rain will bring it, but the fact is that rain falls according to atmospheric conditions.

- A person may believe they are destined for eternal life, but the fact is that all known organisms die.

Belief can be comforting, inspiring, even motivating. But belief without evidence remains belief. Only when tethered to fact does belief rise into knowledge.

2.3 Fact vs. Opinion

Opinion is a judgment of preference. Fact is a recognition of reality.

- "Chocolate tastes better than vanilla" is opinion.
- "Chocolate contains more theobromine than vanilla" is fact.

Opinions can be debated, shaped by culture, mood, or personality. Facts are not negotiable. Where opinions divide, facts unite.

2.4 Fact vs. Illusion

Illusion is the appearance of reality without the substance of it. A mirage in the desert looks like water, but no water exists there. A magic trick may appear to defy physics, but the laws of reality remain unchanged.

Illusions deceive precisely because they mimic facts. But unlike facts, illusions collapse under scrutiny.

Veritism demands vigilance: never confuse the appearance of a fact with the reality of one.

2.5 The Demonstrability of Facts

The power of a fact lies in its demonstrability. A claim becomes fact only when it can be shown to others, repeated, and tested.

- If only one person sees a ghost, it is testimony.

- If a ghost can be photographed, measured, and tested under controlled conditions, then it becomes a fact.

This is why science advances knowledge: it demands demonstration, not mere assertion. A fact is never private. It is always public, always available to scrutiny.

2.6 Historical Lessons in Facts

The history of humanity is the history of facts slowly triumphing over illusion.

- For centuries, people believed the Earth was flat. The fact of its curvature revealed itself through navigation and astronomy.

- People believed diseases were caused by curses or evil spirits. The fact of germs overturned this belief, saving millions of lives.

- People believed the heavens were fixed and unchanging. The fact of exploding stars revealed the cosmos to be alive and violent.

Each step forward was not faith, not opinion, but the recognition of fact.

2.7 Why Facts Matter

Facts are the only solid ground on which to build:

- In science, facts allow us to build medicine, technology, and knowledge.
- In justice, facts protect the innocent and convict the guilty.
- In daily life, facts keep us from walking into traffic, eating poison, or falling for lies.

When facts are denied, disaster follows. When they are honored, progress flourishes.

2.8 Counterarguments and Confusions

There are those who argue that "facts don't exist," that all knowledge is relative or subjective. But this collapses under its own weight. To claim "there are no facts" is to assert a fact.

Others say "facts change," pointing to shifting scientific knowledge. But facts themselves do not change — only our understanding of them. The Earth did not become round when Copernicus described it. It was always round. Science did not create facts; it uncovered them.

Still others argue that "everyone has their own truth." This is confusion. Everyone may have their own perspective, but truth belongs to reality. A person may feel deeply that the Earth is flat, but the fact remains otherwise.

2.9 The Moral Weight of Facts

To ignore a fact is not merely an intellectual error — it is an ethical failure. To act against reality knowingly is to lie to oneself and to others. The one who denies facts corrodes trust, undermines knowledge, and endangers all who depend on truth.

Facts carry a moral weight because they are the ground of integrity. A person who builds on them can be trusted. A person who denies them cannot.

2.10 The Second Principle of Veritism

Thus we arrive at the second principle of this philosophy: Facts are the measure of truth.

All claims must be tested against reality. All beliefs must be judged by evidence. All opinions must yield when confronted by demonstrable fact.

Without facts, there is no truth. Without truth, there is no knowledge. Without knowledge, there is no progress.

To deny facts is to deny reality itself. To embrace them is to align with existence.

Chapter 3 — Truth as Correspondence

Truth is not a matter of taste. It is not created by opinion, secured by belief, or ratified by the number of people who agree with it. Truth is a reflection of reality, nothing more and nothing less.

To say something is true is to say: it corresponds with what is.

This simple recognition — truth as correspondence — is the third cornerstone of Veritism.

3.1 What Is Truth?

Truth is the alignment of thought with reality.

- To say "snow is cold" is true because it corresponds with the fact of snow's temperature.

- To say "the Earth is round" is true because it corresponds with the observed shape of the planet.

- To say "two plus two equals four" is true because it corresponds with the logic of mathematics and the reality of quantity.

A statement is not true because we wish it, because we feel it, or because we agree upon it. A statement is true only when it matches the facts of existence.

3.2 The Myth of Relative Truth

One of the most corrosive illusions of our time is the idea that "truth is relative." People speak of "my truth" and "your truth," as though truth were a possession, a style, or an accessory. But reality does not fracture itself into personal versions.

- If one person says the Earth is flat and another says it is round, both cannot be true.

- If one person says smoking causes cancer and another denies it, the fact does not shift with opinion.

- If one culture believes lightning is divine and another believes it is natural, the lightning itself does not change.

There are not multiple truths — there is one reality. Opinions about it may differ, but truth itself remains singular.

3.3 The Social Function of Truth

Truth matters not only for philosophy but for survival. If truth were relative, communication would collapse. Imagine two builders, one holding that "a meter is 100 centimeters" and the other that "a meter is whatever feels right." They could never build a bridge.

Civilization rests on correspondence truth. Medicine, engineering, science, and law all demand alignment with reality. Without it, planes fall from the sky, contracts dissolve, and justice perishes.

Relativism is comfortable in conversation, but catastrophic in practice.

3.4 Belief vs. Truth

Belief is an internal conviction. Truth is external alignment. The two may coincide, but belief does not create truth.

A man may believe with all his heart that he can walk through a wall. But when he tries, the wall will prove otherwise. Belief without correspondence is delusion.

Veritism makes this demand: do not ask whether you believe something, ask whether it is true.

3.5 Consensus vs. Truth

Truth is not determined by majority vote. Many once believed the Earth was the center of the universe. Their agreement did not make it so.

Consensus can be a useful guide — if many people agree on something, it often indicates a pattern of evidence. But consensus without evidence is only collective belief. History shows again and again that one fact can overturn a thousand opinions.

Truth belongs to reality, not to the crowd.

3.6 Power vs. Truth

Truth is not defined by authority. Kings, priests, governments, and ideologues have all declared themselves the guardians of truth. But when their declarations contradict reality, they crumble.

- Galileo was threatened by the church, yet the Earth continued to orbit the Sun.

- Dictators have proclaimed economic fantasies, yet their nations starved.

- Leaders have declared wars won while reality buried their soldiers.

Power can silence people, but it cannot silence reality. Truth is the one thing that resists authority absolutely.

3.7 The Practical Test of Truth

How, then, do we know what is true? Through evidence and correspondence.

- If a claim matches observable reality, it is true.

- If a claim fails to match, it is false.

- If evidence is incomplete, the truth is not yet known.

This test applies universally. It does not matter whether the claim is scientific, political, religious, or personal. The measure is always the same: does it correspond with reality?

3.8 The Ethical Dimension of Truth

To deny truth is not just an error; it is a betrayal. When a person knowingly denies facts, they place illusion above reality and endanger everyone who depends on honesty.

Lies corrode trust. Denial erodes responsibility. Societies that abandon truth collapse into chaos, because without correspondence, there can be no justice, no cooperation, no progress.

Truth is more than intellectual — it is moral. To honor truth is to honor reality. To deny truth is to betray existence.

3.9 Counterarguments and Their Failure

Some claim that "truth is unknowable." It is true that human knowledge is limited — but limitation does not mean impossibility. We may not know everything, but we know enough to distinguish reality from illusion. To say "truth is unknowable" is itself a truth claim — and thus self-contradictory.

Others argue that "truth is constructed by language." But language only names truth; it does not create it. The word "tree" may differ across cultures, but the tree itself remains unchanged.

Still others argue that "truth changes." But truth does not change — only our recognition of it does. The Earth did not shift from flat to round. It was always round. What changed was human understanding.

3.10 The Third Principle of Veritism

Thus we arrive at the third principle of this philosophy: Truth is correspondence with reality.

- Not belief.
- Not opinion.
- Not consensus.

- Not authority.

Truth belongs to existence. It is the alignment of thought with what is.

To live in truth is to live in reality. To live outside of it is to wander in illusion.

Part II: Knowledge, Reason and Evidence

Chapter 4 — Evidence: The Measure of Truth

If truth is the correspondence between thought and reality, then evidence is the yardstick by which we measure that correspondence. Evidence is not decoration to an argument. It is not a courtesy. It is the very means by which truth is distinguished from error, reality from illusion.

Without evidence, claims dissolve into noise. With evidence, truth becomes knowable.

4.1 What Is Evidence?

Evidence is the trace reality leaves behind when it is observed. It is the imprint of the real upon our senses, our instruments, and our reason.

- Footprints in the snow are evidence of a traveler.

- Fossils in rock are evidence of ancient life.
- Data from a telescope is evidence of distant stars.
- A blood test is evidence of health or illness.

Evidence is reality's own testimony. It is the bridge between what exists and what we know.

4.2 The Necessity of Evidence

Every claim divides into two categories:

1. Supported by evidence → possible candidate for truth.
2. Unsupported by evidence → mere assertion.

No amount of passion, conviction, or repetition can substitute for evidence. A man may shout "the Earth is flat" a thousand times; without evidence, his claim is weightless.

Veritism demands this discipline: before you believe, before you act, before you build, ask — what is the evidence?

4.3 Types of Evidence

Evidence appears in different forms, each suited to different kinds of truth.

- Empirical evidence — derived from sensory observation. Example: water boils at 100°C under standard pressure.

- Logical evidence — derived from reason itself. Example: if all humans are mortal, and Socrates is human, then Socrates is mortal.

- Testimonial evidence — derived from witness reports, valuable when verified and corroborated.

- Scientific evidence — derived from controlled experiment and reproducible data.
- Historical evidence — derived from records, artifacts, and documents, weighed against context.

Each type has its own reliability and limits, but all serve one master: correspondence with reality.

4.4 The Hierarchy of Evidence

Not all evidence is equal. Veritism establishes a hierarchy:

1. Direct observation — seeing the sun rise.
2. Instrumental measurement — telescope data of the sun's spectrum.
3. Independent replication — multiple observers, same results.

4. Consilient evidence — many lines of evidence converging on one fact (e.g., evolution supported by genetics, fossils, and observation).

5. Testimony without verification — weakest, though not worthless, if corroborated.

The higher the evidence on this scale, the stronger its claim to truth.

4.5 Evidence and Probability

Evidence does not always yield certainty; often it yields probability.

- A fingerprint at a crime scene is strong evidence of presence, but not absolute proof of guilt.

- Clouds on the horizon are evidence of rain, but not certainty of a storm.

Truth is sometimes grasped in degrees. Veritism accepts this: knowledge need not be perfect to be valid. Probability, when measured by evidence, is superior to certainty declared without it.

4.6 The Abuse of Evidence

Evidence can be ignored, twisted, or fabricated. The enemies of truth use these tactics:

- Cherry-picking — selecting only evidence that supports a claim while discarding the rest.

- Confirmation bias — seeking only evidence that confirms one's belief.

- Fabrication — inventing false evidence to deceive.

- Misinterpretation — drawing conclusions the evidence does not justify.

Veritism counters these abuses with discipline: examine all evidence, consider alternatives, test assumptions, and follow where reality leads — even against desire.

4.7 Absence of Evidence vs. Evidence of Absence

A common confusion must be addressed. "Absence of evidence is not evidence of absence." True — but with limits.

- If a search is shallow or incomplete, lack of evidence proves little.
- But if a thorough search finds nothing where evidence should exist, absence becomes evidence.

Example: If someone claims there is an elephant in the room, and no one sees, hears, or touches it, absence of evidence is evidence of absence.

Veritism applies this principle with care, always asking: has the search been sufficient to expect evidence?

4.8 Evidence and Belief

Belief without evidence is indistinguishable from fantasy.

To claim something without evidence is to ask others to accept illusion on trust. To believe without evidence is to live in darkness. To act without evidence is to gamble blindly with reality.

The Veritist does not demand certainty before acting — but he demands evidence proportional to the claim. Extraordinary claims require extraordinary evidence. Small claims require modest evidence. In all cases, evidence must guide belief.

4.9 The Ethical Duty of Evidence

Truth-seeking is not only intellectual but ethical. To ignore evidence is to lie, not only to others but to oneself. To falsify evidence is to commit treason against reality.

Every court of justice rests on this principle. Every scientific breakthrough depends on it. Every honest relationship demands it. Without evidence, trust collapses, progress halts, and societies decay.

To honor evidence is to honor reality itself.

4.10 The Fourth Principle of Veritism

Thus the philosophy stands upon its fourth principle: Evidence is the measure of truth.

- Without evidence, there can be no knowledge.

- Without evidence, truth cannot be tested.

- Without evidence, reality is abandoned for illusion.

Evidence is not optional. It is the foundation of reason, the compass of truth, and the safeguard of freedom.

Chapter 5 — Reason: The Tool of Comprehension

Reality exists. Facts are its structure. Truth is the correspondence of thought with reality. Evidence measures this correspondence. Yet without a means of integrating evidence into knowledge, humanity would remain blind, scattered, and lost. That means is reason.

Reason is the compass of the mind. It is the faculty that takes evidence, organizes it, tests it, and draws from it conclusions that can guide action. Without reason, evidence lies dormant, facts remain meaningless, and truth remains undiscovered.

5.1 What Is Reason?

Reason is the disciplined use of logic to connect evidence with truth. It is the process of identifying patterns, eliminating contradictions, and arriving at conclusions that reflect reality.

- Evidence gives us what is observed.
- Reason tells us what it means.

To see smoke is evidence. To conclude there is fire nearby is reason.

Reason transforms raw perception into knowledge.

5.2 The Distinction Between Reason and Instinct

Animals survive by instinct. They respond automatically to their environment. Humans, by contrast, survive by thought. A bird builds its nest the same way every time; a human designs buildings that reach the sky.

Instinct may serve survival, but it does not create understanding. Reason is the uniquely human power to abstract, infer, and project — to grasp not only the immediate, but the universal.

5.3 The Logic of Non-Contradiction

At the root of reason lies a fundamental law: a thing cannot both be and not be in the same respect at the same time.

This is the law of non-contradiction.

- If water is boiling, it cannot simultaneously not be boiling.
- If two plus two equals four, it cannot simultaneously equal five.

Contradictions signal error. Where there is contradiction, there is falsehood. Reason demands consistency because reality is consistent.

5.4 Deduction and Induction

Reason operates through two primary modes:

- Deduction — moving from general principles to specific conclusions. Example: All metals expand when heated. Iron is a metal. Therefore, iron expands when heated.

- Induction — moving from specific observations to general principles. Example: Every swan observed so far is white. Therefore, swans are generally white.

Deduction guarantees truth when the premises are true. Induction provides probability, expanding knowledge through patterns of evidence. Together, they form the twin engines of reason.

5.5 Reason vs. Emotion

Emotion is not a tool of knowledge. Feelings may signal value or danger, but they do not establish truth.

- A man may feel convinced that someone dislikes him, but without evidence, the feeling is mere suspicion.

- A woman may feel hopeful that her illness will vanish, but without treatment, reality may prove otherwise.

Emotion has its place in human life — it motivates, enriches, and colors experience. But in the search for truth, emotion must yield to reason. To confuse the two is to mistake the compass of truth for the weather of mood.

5.6 Reason and Faith

Faith is belief without evidence, often against reason. Reason and faith are not partners; they are opposites.

- Reason asks: What is the evidence?
- Faith says: No evidence is needed.

History shows the results:

- When medicine follows faith over reason, people die.

- When science bows to faith, discovery halts.

- When law submits to faith, justice collapses.

Veritism recognizes reason as the sole valid tool of comprehension. Faith has no claim to knowledge.

5.7 The Dangers of Irrationality

When reason is abandoned, anything goes. Conspiracy theories flourish. Superstition reigns. Dictators exploit blind belief. Entire societies collapse into chaos.

The irrational person demands that reality conform to desire rather than aligning desire with reality. Such a stance is doomed. Reality does not bend. The irrational man does.

5.8 The Discipline of Rational Thinking

Reason is not automatic; it must be cultivated. The Veritist commits to the following practices:

1. Question assumptions.
2. Demand evidence.
3. Check for contradictions.
4. Consider alternatives.
5. Follow logic to its conclusion, even against desire.

Rational thinking is not sterile; it is liberating. It frees the mind from illusion and anchors it in reality.

5.9 The Creative Power of Reason

Reason is not merely defensive — it is creative. It builds bridges, cures diseases, writes symphonies, and explores galaxies. By integrating evidence into knowledge, and knowledge into action, reason expands the horizon of human possibility.

Where faith clings to tradition, reason creates innovation. Where emotion falters in crisis, reason plots solutions. Where instinct stops at survival, reason achieves flourishing.

5.10 The Fifth Principle of Veritism

Thus we arrive at the fifth pillar of this philosophy: Reason is the tool of comprehension.

- Without reason, evidence is inert.
- Without reason, truth cannot be grasped.
- Without reason, human life collapses into chaos.

Reason is humanity's light in the darkness. To abandon it is to extinguish the only flame that makes knowledge possible.

Chapter 6 — Knowledge: The Integration of Truth

Reality provides the ground. Facts mark its structure. Truth corresponds to those facts. Evidence measures the truth of claims. Reason processes the evidence. But what emerges from this process is something greater than its parts: knowledge.

Knowledge is not scattered fragments of data. It is the integration of truths into a coherent system of understanding. It is the difference between a pile of bricks and a finished building.

6.1 What Is Knowledge?

Knowledge is justified, true understanding of reality. It is more than memory, more than belief, more than information. Knowledge is evidence tested by reason and integrated into a consistent whole.

- Data is raw observation: numbers, images, fragments.

- Information is organized data: a chart, a record, a list.

- Knowledge is information validated and explained: an understanding of why and how things are.

Knowledge is what turns facts into wisdom, allowing action, prediction, and progress.

6.2 Knowledge as Context

A single fact is meaningless without context.

- The number "98.6" means little until we know it represents average human body temperature.

- A fossil is a rock until placed in the framework of geology and biology.

- A law is ink on paper until it exists within a system of governance.

Knowledge is the framework of meaning. It connects facts so they do not float in isolation but form a web of understanding.

6.3 The Growth of Knowledge

Knowledge is cumulative and self-correcting. Unlike faith or dogma, it does not freeze itself in a final revelation. It expands, revises, and refines as evidence grows.

- Astronomy grew from the naked eye to telescopes to space probes.
- Medicine grew from herbs to anatomy to genetics.
- Physics grew from Aristotle's categories to Newton's mechanics to Einstein's relativity to quantum theory.

Each new layer does not destroy knowledge but refines it, placing old truths in deeper context.

6.4 The Integration of Disciplines

Knowledge is not limited to one domain. It integrates across fields.

- Biology draws upon chemistry.
- Chemistry rests upon physics.
- Physics requires mathematics.
- History demands archaeology, linguistics, and anthropology.

All knowledge ultimately converges because reality is one. The unity of knowledge reflects the unity of existence.

6.5 The Difference Between Knowledge and Opinion

Opinion is personal; knowledge is impersonal.

- To say "I think this medicine works" without evidence is opinion.

- To say "This medicine works because clinical trials show it reduces symptoms in 80% of patients" is knowledge.

Opinion may be useful as hypothesis, but only evidence and reason elevate it into knowledge.

6.6 Knowledge and Certainty

Knowledge need not mean absolute certainty. Human understanding is always limited by perspective and tools. But within those limits, knowledge can be reliable, precise, and practical.

- We may not know every detail of climate dynamics, but we know enough to predict warming trends.

- We may not know every cause of cancer, but we know enough to treat and prevent it.

Knowledge does not wait for perfection; it operates on what is known, while remaining open to revision.

6.7 The Fragility of Knowledge

Knowledge is not indestructible. It can be lost, ignored, or corrupted.

- Ancient libraries burned, and centuries of understanding were erased.

- Dogma suppressed inquiry, halting progress for generations.

- Modern misinformation corrodes trust in science and fact.

To preserve knowledge requires constant vigilance: evidence must be recorded, tested, and passed on. The decay of knowledge is always possible, but its renewal is always achievable.

6.8 Knowledge as Power and Responsibility

Knowledge empowers action. To know is to anticipate, to prevent, to build. But knowledge also brings responsibility.

- Nuclear physics enabled both power plants and atomic bombs.
- Genetics enables both cures and manipulation.
- Technology enables both communication and surveillance.

Knowledge must be guided by ethics, or its power can be turned against life itself. The Veritist sees knowledge not as neutral accumulation but as a responsibility toward reality and humanity.

6.9 Collective and Individual Knowledge

Knowledge exists both in the mind and in society.

- Individually, knowledge allows a person to navigate reality.
- Collectively, knowledge builds civilization: schools, libraries, sciences, technologies.

A single person may master much, but no one can master all. Knowledge thrives when shared, tested, and refined across minds.

6.10 The Sixth Principle of Veritism

Thus we arrive at the sixth principle of this philosophy: Knowledge is the integration of truth through evidence and reason.

- Without knowledge, facts remain fragments.
- Without knowledge, truth remains isolated.
- Without knowledge, humanity remains blind to its own potential.

Knowledge is not given; it is built. Not once, but continuously. It is the great human project: to weave the truths of reality into an ever-deepening map of existence.

Part III: Wisdom, Ethics and Freedom

Chapter 7 — Wisdom: The Application of Knowledge

Knowledge by itself is powerful but incomplete. One may know a great deal and yet live foolishly. Wisdom is the art of applying knowledge rightly — turning truth into action, and action into flourishing.

Wisdom is knowledge in motion, guided by judgment. It is not merely to know what is true, but to discern what ought to be done in light of that truth.

7.1 What Is Wisdom?

Wisdom is the alignment of action with knowledge. It is knowledge enriched by perspective, tempered by experience, and oriented toward the good.

- Knowledge says: fire burns.
- Wisdom says: use fire to cook food, not to set homes ablaze.

Wisdom transforms the raw material of knowledge into the architecture of a good life.

7.2 The Distinction Between Knowledge and Wisdom

Knowledge without wisdom can be dangerous. A man may know how to make explosives, but without wisdom, he may destroy himself and others.

- Knowledge answers what is.
- Wisdom answers what to do with it.

The two must be joined, or both are diminished. Knowledge unguided by wisdom becomes recklessness. Wisdom without knowledge becomes superstition.

7.3 The Sources of Wisdom

Wisdom arises from several streams:

1. Knowledge — the raw material of reality, tested by evidence and reason.
2. Experience — the lessons of life lived, successes and failures observed.
3. Reflection — the deliberate examination of actions and consequences.
4. Ethics — principles that guide choice in light of value and responsibility.

Together, these sources produce not just knowing, but knowing how to live.

7.4 The Practical Nature of Wisdom

Wisdom is practical by definition. It shows itself in choices, not in abstractions.

- To choose fairness over deceit is wisdom.

- To seek health over indulgence is wisdom.

- To preserve knowledge for the next generation is wisdom.

Wisdom is not passive; it acts. It is the capacity to live well in reality as it is, not as one wishes it to be.

7.5 The Ethical Dimension of Wisdom

Wisdom is inseparable from ethics, because to live well is not only to survive, but to flourish with integrity.

A wise person seeks not only their own good but the harmony of the good with others. Wisdom recognizes that actions ripple outward, shaping families, communities, and societies.

Knowledge without ethics produces cruelty; ethics without knowledge produces futility. Wisdom unites them into a single path of responsibility.

7.6 The Test of Time

Wisdom reveals itself over time. Decisions made in haste, guided by impulse or emotion, often collapse. Decisions made in wisdom endure.

- A wise leader's policies strengthen generations beyond their lifetime.
- A wise builder's structures stand for centuries.
- A wise thinker's principles guide civilizations long after their voice is gone.

Wisdom is tested not by momentary satisfaction, but by lasting consequence.

7.7 The Fool and the Wise

The fool confuses desire with truth. The fool mistakes confidence for competence. The fool believes knowledge alone is enough.

The wise, by contrast, remain humble before reality. They recognize limits, weigh risks, and judge in proportion to evidence. The wise know that ignorance exists, and they account for it.

Wisdom is not arrogance in knowledge, but humility in application.

7.8 Collective Wisdom

Societies, like individuals, can act wisely or foolishly.

- A wise society invests in education, science, and justice.
- A foolish society suppresses knowledge, rewards corruption, and consumes its own future.

Collective wisdom emerges when knowledge, reason, and ethics guide institutions. Collective folly arises when ignorance, fear, and power dominate. History records both paths.

7.9 The Pursuit of Wisdom

Wisdom is not a gift but a pursuit. It requires deliberate effort:

- To learn continuously.
- To reflect on actions.
- To listen to experience — one's own and others'.
- To align choices with evidence, truth, and ethics.

The pursuit of wisdom is lifelong, because reality is inexhaustible and life unendingly complex.

7.10 The Seventh Principle of Veritism

Thus we arrive at the seventh pillar of this philosophy: Wisdom is the application of knowledge toward flourishing.

- Without wisdom, knowledge is dangerous.
- Without knowledge, wisdom is blind.
- Together, they guide human life in harmony with reality.

Wisdom is the crown of understanding, the bridge from knowing to living. It is not only the possession of truth, but the practice of truth.

Chapter 8 — Ethics: Living in Alignment with Reality

Wisdom applies knowledge to the whole of life. But when knowledge touches the question of how we treat ourselves and others, it becomes ethics. Ethics is the branch of wisdom that governs choice and action, aligning human conduct with reality, truth, and value.

To live ethically is to live in truth — to act in accordance with reality, rather than against it.

8.1 What Is Ethics?

Ethics is not a set of arbitrary rules, nor merely the customs of a given culture. Ethics is the recognition that human actions have consequences — and that those consequences can be measured against truth and reality.

An ethical system seeks to answer:

- What should I do?
- How should I live?
- What is the good life?

In this sense, ethics is the science of conduct.

8.2 The Root of Ethics in Reality

All living beings act according to their nature. A tree grows toward the sun; an animal hunts or grazes to survive. Humans, however, possess reason — the ability to choose their actions with knowledge. With choice comes responsibility.

Ethics arises because human beings can act in ways either consistent or inconsistent with reality. To live well, we must act in harmony with the facts of existence.

For example:

- Ignoring the reality of health leads to illness.
- Ignoring the reality of justice leads to conflict.
- Ignoring the reality of truth leads to collapse.

Ethics is the discipline of aligning action with reality.

8.3 The False Roots of Ethics

For centuries, people have sought the foundation of ethics in authority, tradition, or divine decree. But these roots are unstable:

- Authority may be corrupt.
- Tradition may be misguided.
- Divine decree may be unverifiable.

If ethics is to be universal and unshakable, it must rest not on command but on reality itself.

Veritism holds: ethics is not given from above, nor invented at whim. It is discovered, like truth itself, in the structure of reality.

8.4 The Principle of Human Flourishing

The measure of ethics is human flourishing. To flourish is to live in health, strength, knowledge, creativity, freedom, and harmony. Flourishing requires alignment with truth at every level: physical, intellectual, and moral.

- To pursue health is ethical, because it sustains life.

- To pursue knowledge is ethical, because it clarifies truth.

- To pursue justice is ethical, because it sustains society.

Ethics is not self-denial, but self-realization — the fulfillment of life in harmony with reality.

8.5 Responsibility and Freedom

Freedom is the capacity to choose. But without responsibility, freedom degenerates into chaos. Ethics is what transforms freedom into a creative force rather than a destructive one.

- Freedom without ethics leads to exploitation.
- Ethics without freedom leads to tyranny.
- The union of both produces dignity.

To live ethically is to exercise freedom in recognition of responsibility — responsibility to oneself, to others, and to truth.

8.6 Individual Ethics

On the personal level, ethics requires honesty, integrity, and courage.

- Honesty — living by truth rather than falsehood.
- Integrity — consistency of action with principle.
- Courage — the willingness to stand by truth even against resistance.

An ethical individual is not flawless but strives to live in truth, recognizing that every act shapes character.

8.7 Social Ethics

Human beings do not live in isolation. Society is the network of relationships through which individuals flourish or suffer. Ethical societies are built upon justice, fairness, and cooperation.

- Justice recognizes the equal reality of all persons.

- Fairness ensures proportionate reward and responsibility.
- Cooperation allows collective flourishing beyond individual limits.

Ethics in society is not about uniformity but about mutual respect — the recognition that each person is a bearer of truth and value.

8.8 The Consequences of Immorality

Immorality is not merely "breaking rules." It is living against reality. The liar undermines trust; the cheat undermines fairness; the tyrant undermines freedom.

The collapse of civilizations often begins with ethical collapse. When truth is abandoned, when corruption spreads, when justice is denied — reality itself becomes the destroyer.

Immorality is unsustainable because it builds upon falsehood. Sooner or later, reality asserts itself.

8.9 Toward a Universal Ethics

A true system of ethics must be universal — applying to all people, not just some. This does not mean every culture is identical, but that the underlying principles are rooted in reality, which is universal.

- Truth is universal.
- Evidence is universal.
- Reason is universal.
- So too must be ethics, if it is to endure.

Veritism therefore holds: ethics is not relative to opinion but relative to reality. Just as truth does not change with belief, neither does ethics.

8.10 The Eighth Principle of Veritism

Thus emerges the eighth principle: Ethics is the practice of living in alignment with reality, truth, and responsibility, in pursuit of human flourishing.

Ethics is not an external command, but an internal compass, drawn from the very structure of existence. It is the wisdom of living well, in harmony with truth, for the sake of life.

Chapter 9 — Freedom: The Power to Live by Truth

Freedom is not merely the absence of chains. It is the power to live by truth, to act in alignment with reality without coercion, deception, or domination. To be free is to be able to choose, to act, and to bear responsibility for those choices.

Without freedom, truth is silenced, reason is stifled, and ethics becomes impossible.

9.1 The Essence of Freedom

Freedom is not license to do whatever one wishes. It is the condition that allows the individual to act in accordance with reality. True freedom is inseparable from truth.

A man who lives in delusion may believe he is free, but he is a prisoner of falsehood. A society that suppresses truth may claim to grant freedom, but it binds its citizens in invisible chains.

Freedom is not chaos — it is order aligned with reality.

9.2 Freedom and Responsibility

Freedom without responsibility is self-destruction. Responsibility without freedom is slavery. The two must coexist.

- Responsibility grounds freedom in reality.
- Freedom empowers responsibility with choice.

A society that seeks freedom without responsibility descends into corruption. A society that seeks responsibility without freedom collapses into tyranny. Only their union allows human beings to flourish.

9.3 The Enemies of Freedom

Throughout history, freedom has faced three main enemies:

1. Tyranny — the rule of force, where power dictates truth.
2. Dogma — the rule of belief, where faith suppresses evidence.
3. Deception — the rule of falsehood, where lies enslave the mind.

Tyranny enslaves the body. Dogma enslaves the mind. Deception enslaves both.

The defense of freedom is therefore inseparable from the defense of truth.

9.4 Freedom and Truth

Freedom depends on truth, because only in truth can choices be real. A man given false information cannot make a free choice; he acts under illusion. A people fed lies cannot exercise freedom; they live in manipulation.

Thus, truth is not only an intellectual necessity but a moral one. A society that hides truth robs its people of freedom. A society that honors truth empowers its people to live freely.

9.5 Freedom and Reason

Freedom and reason are inseparable. Where freedom is denied, reason is silenced. Where reason is suppressed, freedom is lost.

Reason thrives only when individuals may question, test, and speak without fear. Freedom thrives only when individuals are guided by reason rather than blind force.

The suppression of reason is always the prelude to the suppression of freedom.

9.6 Individual Freedom

For the individual, freedom means self-direction: the ability to guide one's own life in accordance with knowledge and truth.

- The free individual thinks for himself, not as dictated by authority.
- The free individual acts with responsibility, not as driven by whim.
- The free individual builds his life upon reality, not illusion.

Freedom begins not in politics but in the mind. A man enslaved by fear, addiction, or delusion is not free, no matter his outward circumstances.

9.7 Social Freedom

For society, freedom requires justice, protection of rights, and respect for the dignity of every person. Social freedom is not the absence of law but the presence of laws that defend truth, fairness, and responsibility.

A just society does not grant freedom as a gift, nor revoke it as a punishment. It recognizes freedom as the natural condition of all human beings, grounded in their capacity for truth and reason.

9.8 The Fragility of Freedom

Freedom is rare and fragile. History shows that it is easily lost, often surrendered not to conquerors but to comfort. People trade freedom for promises of security, stability, or ease. Yet the loss of freedom always comes at a greater cost: the loss of truth, dignity, and progress.

Eternal vigilance is required, not only against tyrants but against the temptation of false comforts that erode freedom from within.

9.9 Freedom as the Power to Live by Truth

In Veritism, freedom is defined not as the power to do whatever one pleases, but as the power to live by truth.

- A free life is one guided by facts rather than illusions.
- A free mind is one open to reason rather than dogma.
- A free society is one built upon truth rather than tyranny.

Freedom is the condition in which ethics becomes possible, responsibility becomes real, and flourishing becomes achievable.

9.10 The Ninth Principle of Veritism

Thus emerges the ninth principle: Freedom is the power to live by truth, exercised with responsibility, safeguarded by justice, and sustained by reason.

Freedom is not granted by rulers or traditions. It is discovered in reality, defended by truth, and lived by the courage of individuals who refuse illusion.

Part IV: Purpose Living in the Real World

Chapter 10 — Purpose: The Meaning of Life in a Real World

Purpose is the compass of existence. It gives direction to freedom, substance to ethics, and coherence to life. Without purpose, knowledge lies unused, wisdom fades into passivity, and freedom dissolves into wandering.

To live with purpose is to live with clarity, with direction rooted in reality. Purpose transforms mere survival into flourishing, and turns existence into a life well-lived.

10.1 What Is Purpose?

Purpose is the conscious choice of direction in life. It is the deliberate answer to the question: What am I living for?

Purpose is not given by the heavens, decreed by tradition, or imposed by rulers. It is discovered in reality and chosen by reason. Each individual must confront this question honestly and shape their life accordingly.

10.2 The False Sources of Purpose

Throughout history, many have sought purpose in illusions:

- In promises of an afterlife rather than the realities of this life.
- In obedience to authority rather than the freedom of the mind.
- In arbitrary social roles rather than authentic self-direction.

But these false sources enslave rather than liberate. They offer comfort at the cost of truth, and meaning at the cost of integrity. Purpose grounded in illusion is fragile; it collapses when confronted by reality.

10.3 Purpose and Reality

True purpose must be rooted in reality. It must correspond to what is, not merely to what is wished.

- The farmer finds purpose in cultivating the earth, because the earth yields food.

- The scientist finds purpose in discovery, because reality offers knowledge.

- The parent finds purpose in nurturing children, because life continues through them.

Purpose arises not from fantasy but from engaging with the real. To live in truth is to discover purposes that endure.

10.4 The Levels of Purpose

Purpose operates on different levels, each building on the other:

1. Immediate Purpose — the goals of daily life: work, study, care, and creation.

2. Personal Purpose — the shaping of one's character, relationships, and chosen path.

3. Universal Purpose — the recognition that each life participates in the larger reality of existence, contributing to the ongoing story of humanity.

Each level is grounded in facts. Together, they weave the meaning of life in a real world.

10.5 The Relationship Between Purpose and Freedom

Freedom without purpose is drifting. Purpose without freedom is slavery. It is only when the two unite that life becomes fully human.

- A free mind can pursue its chosen purpose.
- A purposeful mind can direct its freedom toward flourishing.

Freedom gives us the power to act. Purpose gives us the reason to act.

10.6 The Role of Knowledge and Wisdom in Purpose

Knowledge and wisdom equip us to choose purpose wisely. Knowledge shows what is possible; wisdom shows what is valuable. Purpose integrates both into action.

A purpose chosen without knowledge leads to failure. A purpose chosen without wisdom leads to regret. But when both unite, purpose becomes the fullest expression of truth in action.

10.7 Purpose and Mortality

The recognition of mortality sharpens the question of purpose. Life is finite; time is limited. Illusions often seek to evade this fact with promises of eternity. But Veritism faces mortality directly and draws strength from it.

To know that life is finite is not despair but urgency. It compels us to live fully, to waste no time on illusions, and to dedicate ourselves to what truly matters. Meaning is found not in endless existence, but in authentic existence.

10.8 Purpose as Contribution

Though each purpose is chosen individually, no life exists in isolation. Purpose finds its fullest expression in contribution — to others, to knowledge, to creation, to the future.

- The teacher contributes by passing on truth.

- The artist contributes by revealing beauty.
- The builder contributes by shaping the world.

Contribution anchors purpose beyond the self, making each life part of something larger without resorting to fantasy.

10.9 The Fulfillment of Purpose

A life lived with purpose is not free of struggle, but it is rich in meaning. Fulfillment arises when action, truth, and value converge in alignment with reality.

Fulfillment is not pleasure alone, nor achievement alone, but the harmony of knowing one has lived authentically, responsibly, and truthfully.

10.10 The Tenth Principle of Veritism

Thus emerges the tenth principle: Purpose is the chosen direction of life, grounded in reality, guided by truth, and fulfilled in contribution to existence.

Purpose is not given from beyond. It is discovered in the real, chosen by the free, and lived through integrity. It is the power that transforms existence into meaning.

Chapter 11 — Community: Shared Reality and Collective Flourishing

No human being exists alone. From birth to death, we are shaped by families, societies, and cultures. Even the most independent thinker relies on the discoveries, tools, and achievements of others. Community is the shared space of human existence — the environment in which individuals live, act, and flourish together.

Veritism holds that true community is not built on illusions, coercion, or conformity. It is built upon shared reality, mutual respect, and the pursuit of collective flourishing.

11.1 The Nature of Community

Community is not merely the gathering of individuals in one place. It is the web of relationships, values, and shared realities that bind people together.

A genuine community arises when individuals:

- Recognize the same reality.
- Respect one another as bearers of truth.
- Work together toward mutual flourishing.

A community built on lies or coercion is not a community but a prison. Only truth sustains lasting bonds.

11.2 Shared Reality as the Basis of Community

For community to endure, its members must share reality. This does not mean every opinion must align, but that facts must be respected.

- A society that denies evidence collapses into superstition.

- A society that suppresses truth collapses into tyranny.

- A society that honors truth grows in freedom, justice, and dignity.

Shared reality is the common ground upon which trust, cooperation, and justice are built. Without it, community disintegrates into division and deception.

11.3 The Role of the Individual in Community

The individual is the smallest unit of community. A strong community does not erase individuality but nourishes it.

- Individuals bring creativity, insight, and responsibility.

- Communities provide support, structure, and shared resources.

- Together, they create more than either could alone.

An ethical community respects individual freedom while fostering mutual responsibility. The flourishing of the whole requires the flourishing of each part.

11.4 False Forms of Community

History is full of false communities — groups bound not by truth but by illusion, coercion, or fear.

- Communities built on ideology demand conformity at the expense of reality.

- Communities built on superstition divide people into believers and heretics.

- Communities built on tyranny sacrifice freedom for control.

Such communities are fragile, for they rest on falsehood. They may endure for a time, but eventually collapse when confronted by reality.

11.5 Community and Ethics

Ethics, applied collectively, becomes justice. A just community is one in which individuals are treated fairly, responsibility is shared proportionately, and truth governs law and policy.

Justice is the lifeblood of community. Without it, cooperation dissolves, trust disappears, and division takes root.

A true community does not require uniformity but demands integrity. Its strength lies not in sameness, but in shared commitment to truth.

11.6 The Balance of Freedom and Responsibility

In community, the balance of freedom and responsibility becomes most visible.

- If freedom is denied, community becomes oppressive.

- If responsibility is denied, community becomes chaotic.

- If both are upheld, community becomes strong.

Each individual must contribute honestly to the whole, while the whole must respect the freedom of the individual. This balance is the key to collective flourishing.

11.7 Contribution and Collective Flourishing

True community flourishes when individuals contribute to the good of all. Contribution takes many forms: work, creativity, care, knowledge, and justice.

When each person gives according to their ability and each receives according to their needs, a harmony emerges that no single person could achieve alone. Contribution is not loss but multiplication: by giving, each strengthens the whole of which they are part.

11.8 The Fragility of Community

Like freedom, community is fragile. It can be undermined by lies, division, and corruption. When trust collapses, community dissolves into isolation.

The health of a community depends upon its commitment to truth. Lies corrode, but truth unites. Communities that cherish reality endure. Communities that abandon it disintegrate.

11.9 Toward a Universal Community

The ultimate vision of Veritism is not a tribal or national community, but a universal one: a human community grounded in shared reality and collective flourishing.

- Truth belongs to no nation.
- Evidence belongs to no culture.
- Reality belongs to all.

A universal community honors diversity while transcending division. It recognizes that all humans share the same reality, the same dependence on truth, and the same responsibility to live in harmony with existence.

11.10 The Eleventh Principle of Veritism

Thus emerges the eleventh principle: Community is the shared reality of individuals living in mutual respect, responsibility, and truth, aimed at collective flourishing.

Community is not the erasure of the self, but the extension of the self into a larger whole. It is where truth becomes social, freedom becomes cooperative, and purpose finds contribution.

Chapter 12 — Fulfillment: Living the Whole of Truth

Fulfillment is not a fleeting pleasure or a passing success. It is the deep, sustained sense of living in harmony with reality — of having aligned one's mind, actions, and relationships with truth. It is the wholeness that arises when life itself becomes an expression of what is real.

Veritism teaches that fulfillment is not granted by fate, chance, or supernatural forces. It is earned through the consistent practice of living truthfully: perceiving facts, reasoning clearly, acting ethically, and contributing meaningfully.

12.1 The Meaning of Fulfillment

Fulfillment is often confused with happiness, but they are not the same.

- Happiness can be momentary; fulfillment endures.

- Happiness may come from illusion; fulfillment requires truth.
- Happiness can be lost with circumstances; fulfillment persists because it is rooted in reality.

Fulfillment is not a gift but an achievement: the reward of living in accordance with reality over the course of a life.

12.2 The Path to Fulfillment

The path to fulfillment follows the very structure of Veritism itself:

1. Facts — Acknowledging reality as it is.
2. Truth — Aligning belief with fact.
3. Evidence — Testing belief against reality.
4. Reason — Using thought to integrate evidence.

5. Knowledge — Building a foundation of truth.

6. Wisdom — Applying knowledge ethically.

7. Ethics — Living rightly in alignment with reality.

8. Freedom — Choosing truth freely and responsibly.

9. Purpose — Directing life toward meaningful goals.

10. Community — Flourishing together in shared reality.

Each stage builds upon the last. Fulfillment is the integration of them all — living the whole of truth.

12.3 Wholeness of the Individual

At the personal level, fulfillment is wholeness. It is when thought, action, and value are no longer at war with each other, but harmonize around truth.

A fulfilled person:

- Lives without self-deception.
- Chooses goals consistent with reality.
- Finds joy not in illusion but in creation, contribution, and truth.

Wholeness dissolves inner conflict and creates a unity of self.

12.4 Fulfillment and Freedom

Fulfillment requires freedom. A life lived under coercion may survive, but it cannot flourish. Without the ability to choose truth, individuals cannot achieve the wholeness of living it.

Freedom is not license but responsibility: the responsibility to align with reality, even when illusions tempt. Fulfillment is the fruit of choosing truth consistently.

12.5 Fulfillment in Community

Fulfillment is not only individual but communal. Just as one person flourishes through truth, so too can a community flourish when reality is its foundation.

A community grounded in truth produces:

- Trust instead of suspicion.
- Justice instead of exploitation.
- Flourishing instead of collapse.

Fulfillment expands outward from the individual to the group, from the group to humanity, and from humanity to its relationship with the wider natural world.

12.6 The Enemies of Fulfillment

Fulfillment cannot coexist with falsehood. The greatest enemies of fulfillment are:

- Illusion — The comfort of lies over the discipline of truth.

- Denial — The refusal to face reality as it is.

- Coercion — The suppression of free reason and choice.

- Neglect — The failure to integrate knowledge into life.

A fulfilled life is one that has overcome these enemies not once, but continually.

12.7 The Joy of Fulfillment

Though fulfillment is not identical with happiness, it brings joy. This joy is deeper than pleasure, stronger than circumstance, and more enduring than desire. It is the joy of wholeness: knowing that one's life has not been wasted on falsehood, but lived fully in reality.

This joy is not loud but steady, not fleeting but lasting. It is the quiet strength of a life well-lived.

12.8 Fulfillment and Mortality

Fulfillment gains its full meaning in light of mortality. Every life is finite. To waste it on illusions is to lose it twice: once in living, and again in dying. To live truthfully is to ensure that when life ends, it has been whole, honest, and real.

Fulfillment does not deny death; it completes life in the face of it. It gives dignity to existence by ensuring that it has been lived in truth.

12.9 The Twelfth Principle of Veritism

Thus emerges the twelfth principle: Fulfillment is the integration of truth across the whole of life, lived in harmony with reality, free from illusion, and directed toward flourishing.

Fulfillment is the crown of Veritism. It is the lived expression of every principle — not abstract, but embodied. It is not what we hope for, but what we build, day by day, choice by choice, truth by truth.

Part V: Legacy in Transcendence

Chapter 13 — Legacy: Truth Across Generations

Human life is finite, but truth endures. Each of us inherits a world shaped by those who came before, and we leave behind a world that will be shaped by what we do now. Legacy is not only about personal remembrance; it is about the transmission of truth across generations.

Veritism teaches that the value of a life is not measured solely by what it achieves in the moment, but also by what it contributes to the enduring story of humanity. Legacy is the continuity of truth.

13.1 The Nature of Legacy

Legacy is the lasting impact of one's actions, knowledge, and values. It is what remains when a life has ended, and what shapes those who live after.

- A false legacy spreads illusion and corruption.
- A hollow legacy leaves no trace.
- A true legacy carries forward the power of reality, evidence, and reason.

The measure of a legacy is whether it strengthens the bond between humanity and truth.

13.2 Inheritance and Responsibility

Each generation inherits the achievements and failures of those before it. Science, knowledge, art, justice — these are gifts passed down. War, superstition, oppression — these too are legacies.

Veritism calls for responsibility: to accept the gifts of truth, to correct the errors of falsehood, and to pass forward what is real, just, and life-affirming.

Legacy is not passive inheritance but active stewardship.

13.3 The Fragility of Legacy

Legacy is fragile. Knowledge can be lost, truth suppressed, wisdom forgotten. History shows civilizations rising in truth and collapsing in illusion.

To preserve legacy requires vigilance:

- Recording knowledge with accuracy.
- Teaching evidence over superstition.
- Protecting freedom of thought.
- Passing forward values grounded in reality.

Without vigilance, the chain of truth can break, and future generations are left to rediscover what was once known.

13.4 Legacy Through Knowledge

Knowledge is one of humanity's most enduring legacies. From cave paintings to digital libraries, from oral traditions to scientific journals, each generation builds upon the truths discovered by the last.

To contribute knowledge is to strengthen humanity's bridge to the future. To distort or suppress knowledge is to weaken it.

Every fact, every discovery, every insight passed forward becomes part of the human inheritance.

13.5 Legacy Through Ethics

Ethical living leaves a moral legacy. The choices we make — how we treat others, how we build communities, how we uphold justice — ripple across time.

Unethical systems, once entrenched, can damage generations. But ethical principles, rooted in reality, create enduring frameworks of trust and fairness.

A life lived in truth plants seeds of justice for those yet to come.

13.6 Legacy Through Community

Communities themselves are legacies. The structures we build — families, institutions, nations — either enable or obstruct future flourishing.

A community grounded in truth passes strength to its descendants. A community built on illusion crumbles, leaving ruin.

To strengthen community in the present is to leave a foundation for those who will live tomorrow.

13.7 Personal Legacy

While legacy often takes the form of collective inheritance, each individual shapes it as well. Every act of honesty, every pursuit of truth, every moment of courage contributes to the unfolding story of humanity.

To live truthfully is to leave a personal legacy that endures in the lives of others, even if one's name is forgotten. Legacy is not about fame, but about impact.

13.8 Humanity's Legacy

Beyond individuals and communities, humanity itself creates a legacy. The Earth we leave behind, the knowledge we safeguard, the principles we uphold — all become part of the story of humankind.

If humanity grounds its legacy in truth, the future is open with possibility. If humanity grounds it in illusion, it risks collapse.

Veritism calls for humanity to recognize itself as a custodian of truth, not just for itself but for all future life.

13.9 The Thirteenth Principle of Veritism

Thus emerges the thirteenth principle: Legacy is the transmission of truth across generations, the responsibility to preserve, strengthen, and pass forward the bond between humanity and reality.

Legacy is the bridge between the fleeting span of a life and the enduring arc of history. To live truthfully is not only to flourish oneself, but to give future generations the means to flourish as well.

Chapter 14 — Transcendence: Truth Beyond the Self

Every philosophy must answer the question of what lies beyond the individual, the community, and even the generations of humanity. Transcendence is not escape into illusion or mysticism — it is the recognition that truth belongs to reality itself, not to us alone.

Veritism rejects fantasies of gods or otherworldly realms. Yet it affirms transcendence in a real and profound sense: truth exceeds the self, exceeds any single life, exceeds any generation, and binds us to the whole of existence.

14.1 The Nature of Transcendence

To transcend is to go beyond. Transcendence in Veritism means moving beyond personal perspective and recognizing one's place in the vast order of reality.

- The individual is finite, but truth is infinite.
- Life ends, but reality continues.
- Generations pass, but the laws of nature endure.

Transcendence is not detachment from reality but deeper unity with it.

14.2 Reality as the Ultimate Context

All existence unfolds within reality. Stars form and die, civilizations rise and fall, individuals are born and pass away — but the framework of reality remains.

To transcend is to recognize that our lives are woven into this larger order. Our truths, our knowledge, and our actions all find their ultimate meaning in the reality that grounds them.

14.3 The Humility of Transcendence

Transcendence demands humility. No individual, no nation, no species can claim ownership of truth. We participate in truth, but we do not create it.

Humility in the face of truth is not weakness but strength. It anchors us, reminding us that while we are finite, the reality we serve is not.

14.4 The Joy of Transcendence

Transcendence also brings joy — the joy of knowing that our lives are part of something greater than ourselves. To live truthfully is not merely to flourish individually or collectively, but to contribute to the unfolding order of existence itself.

This joy is not mystical; it is grounded. It comes from realizing that every act of truth participates in a reality that outlives us.

14.5 Transcendence and Legacy

Legacy ties into transcendence. What we pass forward becomes part of the ongoing movement of truth through history. Yet transcendence looks beyond even history, asking what it means to live truthfully as part of the cosmos itself.

Our legacy may shape generations, but transcendence roots us in something vaster — the enduring fabric of existence.

14.6 Transcendence Without Illusion

Traditional religions promise transcendence through gods, heavens, or spiritual realms. Veritism offers transcendence without illusion: not escape from reality, but deeper immersion in it.

The laws of physics, the structures of matter, the evolution of life — these are not mystical, yet they are profoundly transcendent. They show us that we belong to something immeasurably greater than ourselves.

14.7 Transcendence Through Contribution

Each life, however small, can participate in transcendence through contribution to truth. To discover, to teach, to build, to live honestly — these are ways of embedding one's existence into the enduring story of reality.

Transcendence is not the erasure of the self but the expansion of the self into alignment with what is greater.

14.8 Mortality and Transcendence

Death brings the question of transcendence into sharp relief. For some, mortality is a source of despair. For Veritism, it is the invitation to transcend.

To live truthfully is to ensure that, even though life ends, it has been part of something eternal: the enduring reality of truth. Mortality does not diminish transcendence; it makes it essential.

14.9 The Fourteenth Principle of Veritism

Thus emerges the fourteenth and final principle: Transcendence is the recognition that truth extends beyond the self, binding us to the whole of reality, and giving our finite lives meaning within the infinite order of existence.

Transcendence is not otherworldly but this-worldly. It is not mystical but factual. It is the final step of Veritism: the understanding that truth is larger than us, but that we belong within it.

14.10 Conclusion: Living Veritism

With transcendence, Veritism reaches its completion. It begins with facts, passes through truth, evidence, reason, knowledge, wisdom, ethics, freedom, purpose, community, fulfillment, and legacy — and ends in transcendence.

The journey of Veritism is the journey of living wholly in reality:

- To see clearly.
- To think honestly.
- To act justly.
- To flourish fully.
- To contribute meaningfully.
- To belong ultimately.

This is not a creed, nor a faith, nor a superstition. It is the recognition of reality itself, lived to its fullest depth.

Chapter 15 — Practice: Living Veritism Daily

Philosophy has no power unless it is lived. Veritism is not a system of abstract thought to be admired from a distance; it is a way of engaging with the world day by day. To live truthfully is to embody the principles of Veritism in thought, action, and community.

This final chapter is not about theory but about practice. It shows how to bring Veritism into daily life, so that truth becomes more than an idea — it becomes a way of being.

15.1 The Discipline of Awareness

The first practice of Veritism is awareness: the daily discipline of seeing reality as it is.

- Pause to observe. Before reacting, notice. What are the facts of this moment?

- Separate perception from interpretation. Ask: what is directly evident, and what is my assumption?

- Seek clarity. Train the mind to resist illusion and distortion.

Awareness is not mystical; it is attentiveness to what is real.

15.2 Practicing Fact-Checking in Life

In a world full of misinformation, illusions, and unchecked assumptions, Veritism calls for everyday fact-checking.

- Ask: What evidence supports this claim?

- Ask: Can this be demonstrated, or is it mere assertion?

- Ask: Am I believing this because it is true, or because it is comfortable?

The discipline of fact-checking is the antidote to deception, both external and internal.

15.3 Reason as Daily Compass

Reason must not remain theoretical. It is a daily compass for navigating choices.

- Weigh alternatives. Use reason to examine outcomes, not just desires.

- Avoid contradictions. If two beliefs conflict, test them against reality.

- Practice integration. Connect knowledge across areas of life to see the whole picture.

Reason is the habit of thinking honestly. It must be exercised like a muscle.

15.4 Building Habits of Truth

Habits form the architecture of daily life. To live Veritism, one must create habits aligned with truth.

Examples:

- Begin each day by grounding yourself in a fact (something real, observable, undeniable).
- Question one assumption daily: what do I take for granted, and does it match reality?
- End each day by reflecting: Did I live closer to truth today, or farther from it?

Small practices create lasting change.

15.5 Veritism in Relationships

Living truthfully also applies to how we treat others.

- Honesty. Speak what is real, not what manipulates.
- Respect. Recognize others as fellow seekers of truth, not tools for illusion.
- Boundaries. Refuse to participate in deception, even when it is socially expected.

Truth builds trust. Illusion destroys it. Every relationship becomes stronger when grounded in reality.

15.6 Veritism at Work

In the realm of work, Veritism calls for integrity.

- Ground decisions in evidence, not convenience.
- Pursue excellence by aligning effort with reality, not shortcuts.

- Treat colleagues and customers with the dignity of truth, not manipulation.

Work becomes meaningful when it is an arena for truth, not merely survival.

15.7 Facing Illusions and Lies

No life is free from illusions — personal, cultural, or systemic. The practice of Veritism is learning how to face them.

- When confronted with a lie, test it against reality.
- When tempted by illusion, ask what it costs you to believe it.
- When society demands conformity to falsehood, choose courage over comfort.

Truth is not always easy, but it is always worth it.

15.8 Veritism in Community

Daily practice is not only personal; it is communal.

- Share truth with others, not as dogma, but as clarity.
- Support communities that value evidence and reason.
- Resist communities that demand blind belief.

Living truthfully in community means fostering environments where reality is honored and illusions cannot thrive.

15.9 Mortality and the Daily Practice of Truth

Living truthfully is especially important in light of mortality. Each day is finite; each choice matters.

To live truthfully is to ensure that one's days are not wasted. Fulfillment, legacy, and transcendence are not achieved in a single act, but in the accumulation of days lived in alignment with reality.

15.10 A Daily Framework for Veritism

To make Veritism practical, here is a simple daily framework:

1. Observe — Begin the day with one fact.
2. Question — Challenge one assumption.
3. Reason — Use logic in one decision.
4. Act — Do one thing aligned with truth.
5. Reflect — End the day by asking: Did I live closer to reality today?

These five practices bring the principles of Veritism into ordinary life, transforming philosophy into a way of living.

15.11 The Fifteenth Principle of Veritism

Thus emerges the fifteenth principle: Practice is the daily discipline of living in alignment with reality, grounding thought, action, and community in truth.

Without practice, philosophy is powerless. With practice, truth becomes life.

15.12 Living the Whole of Veritism

The journey of Veritism is complete, but never finished. Each day is another opportunity to live closer to reality, to reject illusion, to embrace truth.

Living Veritism is not about perfection, but about persistence. It is not about being right always, but about being honest always.

The reward is not only clarity, but wholeness: a life lived in truth, for oneself, for others, and for the generations to come.

Conclusion — The Call of Veritism

Philosophy begins in wonder but must end in practice. We began this journey by asking the oldest of questions: What is real? From that question flowed the search for truth, the nature of facts, the role of evidence, the necessity of reason, the structure of knowledge, the wisdom of application, the ethics of alignment, the power of freedom, the purpose of life, the meaning of community, the wholeness of fulfillment, the permanence of legacy, the transcendence of truth beyond the self, and finally, the discipline of practice.

The system that has emerged — Veritism — is not another creed, not another religion, not another ideology. It is the recognition that reality exists, that facts are knowable, that truth corresponds with what is, and that human beings can and must live in alignment with this reality if they are to flourish.

Veritism is not a theory to be believed. It is a framework to be lived.

1. The Simplicity of Reality

The foundation is simple: reality is what it is. No belief, no denial, no wish can change it. The world is not shaped by desire but by fact. To live in harmony with reality is to free oneself from the endless cycle of illusion.

2. The Courage of Truth

Truth demands courage. It asks us to see clearly even when it is painful, to accept limits when they are undeniable, and to stand firm when illusion tempts us with comfort. To live by truth is not always easy, but it is always liberating.

3. The Discipline of Evidence and Reason

Evidence is the measure of truth. Reason is the tool of comprehension. Together they form the twin pillars of Veritism, guiding the seeker through a world of noise and deception. To deny either is to surrender to falsehood. To live by both is to claim one's dignity as a rational being.

4. The Integration of Life

Knowledge is the integration of truth; wisdom is its application. Ethics is the living of reality in thought and action. Freedom is the power to live authentically by truth. Purpose is the discovery of meaning within the real. Community is the weaving together of shared reality. Fulfillment, legacy, and transcendence are the rewards of living fully aligned with what is.

Each chapter has been a step in a larger arc — the arc of human flourishing through truth.

5. The Responsibility of Veritism

To live Veritism is to carry responsibility. Each person is accountable to reality. Each choice either honors or denies truth. To live falsely is to decay; to live truthfully is to flourish.

This responsibility is not a burden but a calling: the invitation to live fully awake, fully real, fully true.

6. The Fifteen Principles as Compass

The fifteen principles of Veritism are not commandments; they are compass points. They guide, not dictate. They do not impose belief but illuminate reality. They are not dogma but discipline.

To practice Veritism is not to follow rules but to live with integrity: to let truth shape thought, action, and being.

7. The Future of Truth

Human history has been a long struggle between reality and illusion, truth and falsehood, reason and irrationality. That struggle continues. Veritism does not end the struggle, but it provides a framework for clarity. It offers a way to navigate the confusion of our age with a simple commitment: live by truth, and let no illusion enslave you.

Closing Words

We do not need another religion, built on faith without evidence.

We do not need another ideology, promising utopias on the basis of slogans.

And we do not need atheism, which rejects gods but too often stops at denial, leaving no framework for how to live.

What we need is a way of life that honors what is real.

A philosophy that affirms reality, not rejects illusions alone.

A path that grounds truth in evidence, reason, and alignment with the world as it is.

This is Veritism: not faith, not negation, not relativism — but reality itself, lived fully.

This is Veritism:

- Reality as foundation.
- Facts as measure.
- Truth as correspondence.
- Evidence as standard.
- Reason as tool.
- Knowledge as integration.
- Wisdom as application.
- Ethics as alignment.
- Freedom as power.
- Purpose as meaning.
- Community as flourishing.
- Fulfillment as wholeness.
- Legacy as continuity.
- Transcendence as beyond self.
- Practice as daily discipline.

Fifteen principles. One path. A life lived in truth.

The call of Veritism is clear: Live reality. Live truth. Live fully.

Acknowledgments

No philosophy is created in a vacuum. Though Veritism is my original work, it was shaped, sharpened, and inspired by the insights of those who came before me.

I wish to acknowledge Ayn Rand, whose uncompromising defense of reason and reality in her philosophy of Objectivism laid a foundation upon which this work could build. In particular, Leonard Peikoff's book, Objectivism: The Philosophy of Ayn Rand, offered a systematic clarity that demonstrated how rigorous a philosophy could — and should — be.

To Neil Peart, whose lyrical genius proved that words can be both art and argument, poetry and philosophy. His ability to capture the struggles of the human spirit with both honesty and hope continues to inspire me.

To Nathaniel Branden, whose pioneering work in psychology and self-esteem illuminated the essential connection between truth, integrity, and the inner life.

To Anthony Robbins, whose book Awaken the Giant Within showed that philosophy is not merely abstract but can be lived practically, energetically, and with purpose.

To Brian Cox, whose voice and presence as a scientist have brought the wonder of the cosmos to countless minds, reminding us that reality itself is more awe-inspiring than any myth.

And to all thinkers, artists, scientists, and seekers of truth who have dared to ask the difficult questions and confront illusions — your work has helped light the path that Veritism now seeks to make clearer.

Finally, I extend my gratitude to the readers of this book. By choosing to engage with Veritism, you are not only exploring a new philosophy but participating in the timeless human project: the search for truth, lived fully in reality.

Also from the Author

Mastering The Art of Mindful Thinking

The Prince of Pico

@

robtaylorbooks.com

© *Copyright Rob Taylor 2025*

www.ingramcontent.com/pod-product-compliance
Lightning Source LLC
Chambersburg PA
CBHW061220070526
44584CB00029B/3906